# Materials and Motion

Full Option Science System
Developed at
The Lawrence Hall of Science,
University of California, Berkeley
Published and distributed by
Delta Education,
a member of the School Specialty Family

1487695
978-1-62571-427-5
Printing 1 — 8/2015
Standard Printing, Canton, OH

# Table of Contents

# The Story of a Chair

Here is a kind of chair you might
sit on in a park.
Do you know how it was made?

A chair begins as a **tree** in the **forest**.

A lumberjack cuts the trees
in the forest.

A truck driver takes the trees
to the sawmill.
The trees are cut into
boards at the sawmill.

The boards go to the lumberyard.
The woodworker buys **wood** at
the lumberyard.

The woodworker makes the chair
from wood.
And then it's put in the park for
you to sit on!

# Are You an Engineer?

**Engineers** ask questions about materials.

Engineers make **observations**.
Engineers test things to find out
about them.

Engineers **compare** results of tests.
Engineers look at how materials are
the same and different.

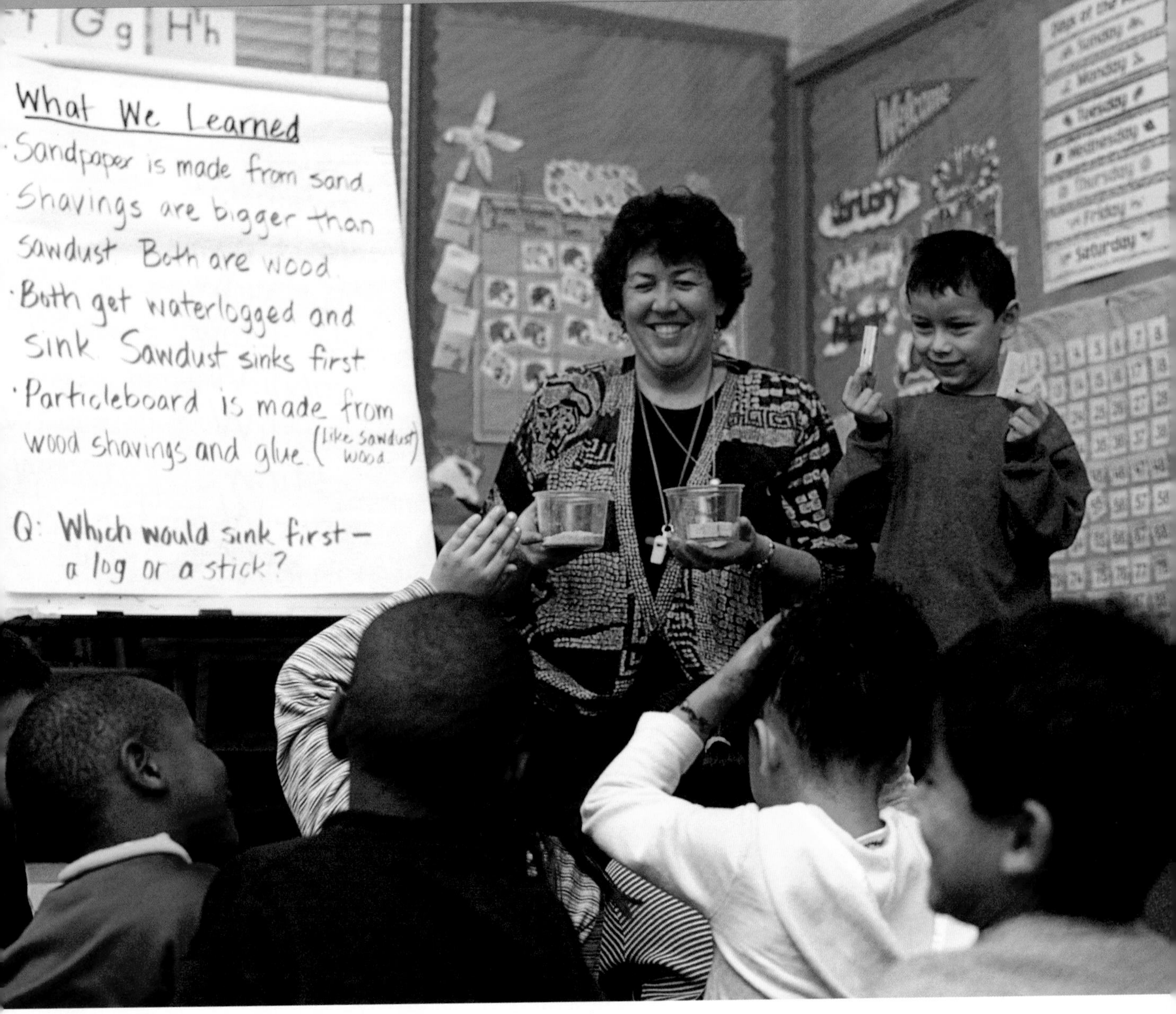
What We Learned
·Sandpaper is made from sand.
Shavings are bigger than
sawdust. Both are wood.
·Both get waterlogged and
sink. Sawdust sinks first.
·Particleboard is made from
wood shavings and glue. (like sawdust wood)
Q: Which would sink first—
a log or a stick?
Welcome

Engineers share their ideas with others.
Engineers improve their designs.

Are you an engineer?

# The Story of a Box

This box is strong.

But it is made of **paper**.

Do you know how a box is made?

A box starts in the forest as a tree.

First, the trees are cut.

Then, the trees go to a paper mill.

The trees are chopped to
make **sawdust**.
Sawdust is soaked in **water**.
This makes **pulp**.

The pulp is flattened and dried
to make paper.
The paper is put on a roll.
This roll of paper can be used
to make cardboard.

The paper is layered and
glued together.
This makes cardboard.
The cardboard is cut and folded.

Now the box is ready to use!

# What Is Fabric Made From?

**Fabric** can be colorful.

Fabric can be nubby or smooth.

Fabric can be strong.

Where does fabric come from?

People make fabric.
Some fabric is made from
animal hair or fur.
Wool comes from sheep.

Wool is spun into yarn.

The yarn is dyed to give it color.

The yarn is woven to make wool fabric.

Silk comes from silkworm cocoons.
Silk fibers are pulled from the cocoons.
The fibers are made into thread.

The thread is woven to make silk fabric.

Some fabric is made from plants.
Cotton comes from cotton plants.
Cotton is spun into thread.

The thread is woven to make cotton fabric.

**Jute** comes from jute plants.
Jute plants are cut and soaked.

Jute is dried and spun into thread. The thread is woven to make burlap fabric.

Not all fabric is made from animals or plants.
Some fabric is made from **oil**.
Oil is pumped from the ground.

Nylon thread is made from oil.
The thread is woven to make
nylon fabric.

Some fabric is woven.

Some fabric is knitted.

Look at the fabric you are wearing.

How do you think it was made?

# How Are Fabrics Used?

Some fabrics are waterproof.

Waterproof fabrics are good for
umbrellas, awnings, and tents.

Some fabrics are easy to wash.

They are good for towels, pants, and shirts.

Some fabrics are colorful.

They are good for flags and
dressing up on special days.

Some fabrics are strong.
They are good for backpacks
and shoes.

Some fabrics are soft.
They keep you warm.
Soft fabrics are good for snuggling!

# What kinds of fabric are you wearing?

# Land, Air, and Water

Do you ever **recycle** paper?
If you do, you are helping to
save a tree.

We use trees for many things.
Trees are important to us.
Other things are important to
us, too.

**Land**, **air**, and water are important.
We need good land for growing food.

We need good water to drink.

We need good air to breathe.
How can we take care of our
land, air, and water?

# I Am Wood

# Pushes and Pulls

Toys don't move by themselves.
How can you move a toy?
You give it a **push**.

A **pull** can move something, too.
A big dog can pull a wagon.

A person can pull a wagon, too.

Something moves only when
pushed or pulled.
A pinwheel can move around
and around.

What is pushing this pinwheel?
The wind pushes the pinwheel around.
What else is the wind pushing?

What moves these two girls down the slide? **Gravity** pulls the girls down the slide.

Gravity pulls things down.

This girl and boy are swinging.
Can you see anyone pushing
or pulling them?
The pull of gravity moves them.

They move one **direction**
and then back.
Gravity helps them
swing back and forth.

**Rolling** is a **motion**.

Round things roll easily.

Marbles, balls, and cans roll easily.

Which things will roll down
the ramp?
What pulls them downhill?

How can you change the
**speed** of a rolling object?
Which ramp makes the
can roll the fastest?
Ready, set, go!

Wheels roll downhill, too.
Gravity makes it easy to
ride a bike downhill.

# Collisions

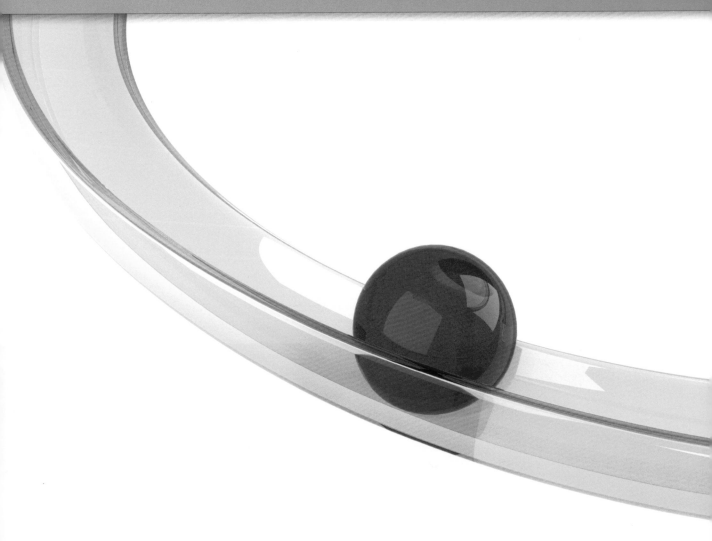

A ball on a **slope** will roll.
Gravity pulls it downhill.
What happens at the bottom
of the slope?

# The ball might **collide** with something.

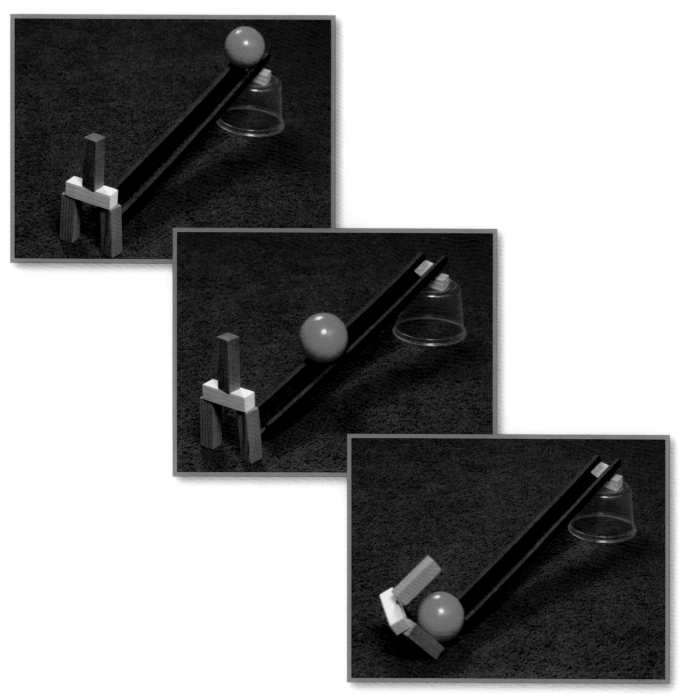

What happens if a moving ball collides with other objects? The ball might push the objects away.

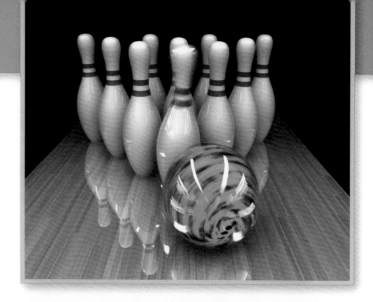

The bowling ball
pushes the pins
away.

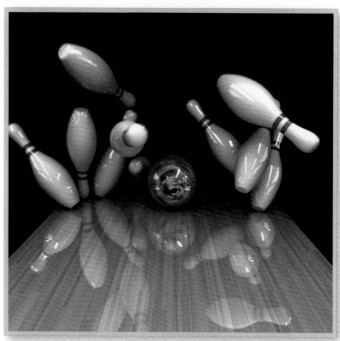

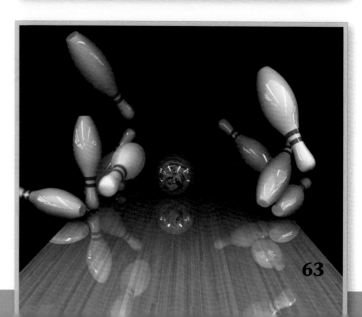

A soccer player gets a pass.
She might kick the ball.
A kick pushes the ball
in a new direction.

A player kicks the ball into the net.
The net pushes on the moving ball.
It stops the ball.

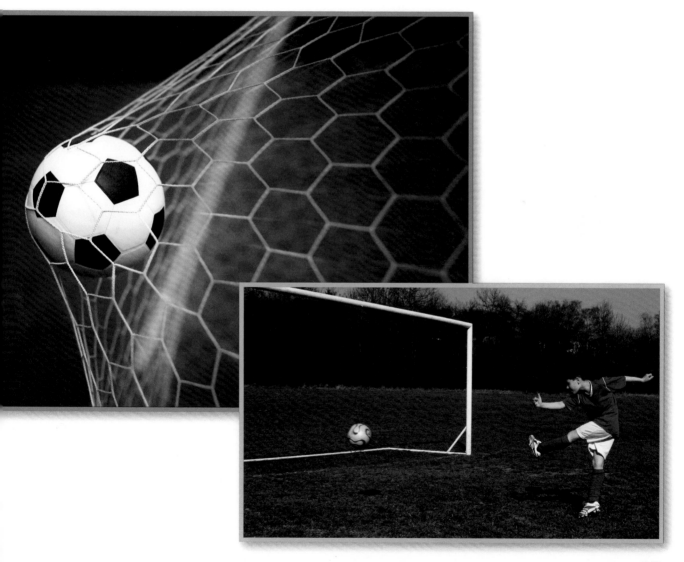

This golfer wants to hit the
ball into a hole.
How can he make the
ball turn a corner?

He might use the walls to push
the ball.
The walls can change the
ball's direction.
The ball rolls into the hole.

This ball got a really big push.
The ball pushed through
a block wall.
That is a big collision!

# Glossary

**air** a mixture of gases that we breathe **(43)**

**collide** to hit or crash **(61)**

**compare** to look at how things are the same and different **(11)**

**direction** the path on which something is moving **(55)**

**engineer** a scientist who designs ways to solve a problem **(9)**

**fabric** a flexible material used to make clothing. Fabric and cloth are the same. **(19)**

**forest** a natural area where trees grow **(4)**

**gravity** a force that pulls things toward Earth **(52)**

**jute** a plant used to make burlap **(26)**

**land** the solid surface of Earth **(43)**

**motion** the act of moving **(56)**

**observation** the act of using your senses to get information **(10)**

**oil** a liquid pumped from the ground. Oil is used by people to make nylon and many other things. **(28)**

**paper** a flexible material made from wood **(13)**

**pull** when you make things move toward you. Pulling is a force. **(48)**

**pulp** a mixture of sawdust and water used to make paper **(15)**

**push** when you make things move away from you. Pushing is a force. **(47)**

**recycle** to use again **(41)**

**rolling** moving from one place to another by going around and around **(56)**

**sawdust** the small bits made when a saw cuts wood **(15)**

**slope** an upward or downward slant **(60)**

**speed** how fast an object moves **(58)**

**tree** a plant with a woody stem, roots, and branches with leaves **(4)**

**water** the most common liquid on Earth. Water is found in oceans, lakes, and streams. **(15)**

**wood** a strong material in the trunks and branches of trees **(7)**